A WE[...]
THE AIRPORT
A HEATHROW DIARY

A WEEK AT THE AIRPORT
A HEATHROW DIARY

ALAIN DE BOTTON
PHOTOGRAPHS BY RICHARD BAKER

P

First published in Great Britain in 2009 by
PROFILE BOOKS LTD
3a Exmouth House
Pine Street
London EC1R 0JH
www.profilebooks.com

10 9 8 7 6 5 4 3 2 1

Photographs © Richard Baker, 2009
Text designed and typeset in ITC Galliard
by Joana Niemeyer at April
Cover design by David Pearson
Printed and bound in Great Britain
by Butler, Tanner and Dennis Ltd, Frome, Somerset

A CIP catalogue record for this book is available from
the British Library.

ISBN 978 1 84668 359 6

Mixed Sources
Product group from well-managed
forests and other controlled sources
www.fsc.org   Cert no. SGS-COC-005091
© 1996 Forest Stewardship Council

For Saul

Contents

1 While punctuality lies at the heart of what we typically understand by a good trip, I have often longed for my plane to be delayed – so that I might be forced to spend a bit more time at the airport. I have rarely shared this aspiration with other people, but in private I have hoped for a hydraulic leak from the undercarriage or a tempest off the Bay of Biscay, a bank of fog in Malpensa or a wildcat strike in the control tower in Málaga (famed in the industry as much for its hot-headed labour relations as for its even-handed command of much of western Mediterranean airspace). On occasion, I have even wished for a delay so severe that I would be offered a meal voucher or, more dramatically, a night at an airline's expense in a giant concrete Kleenex box with unopenable windows, corridors decorated with nostalgic images of propeller planes and foam pillows infused with the distant smells of kerosene.

In the summer of 2009, I received a call from a man who worked for a company that owned airports. It held the keys to Southampton, Aberdeen, Heathrow and Naples, and oversaw the retail operations at Boston Logan and Pittsburgh International. The corporation additionally controlled large pieces of the industrial infrastructure upon which European civilisation relies (yet which we as individuals seldom trouble ourselves about as we use the bathroom in Białystok or drive our rental car to Cádiz): the waste company Cespa, the Polish construction group Budimex and the Spanish toll-road concern Autopista.

My caller explained that his company had lately developed an interest in literature and had taken a decision to invite a writer to spend a week at its newest passenger hub, Terminal 5, situated between the two runways of London's largest airport. This artist, who was sonorously to be referred to as Heathrow's first writer-in-residence, would be asked to conduct an impressionistic survey of the premises and then, in full view of passengers and staff, draw together material for a book at a specially positioned desk in the departures hall between zones D and E.

It seemed astonishing and touching that in our distracted age, literature could have retained sufficient prestige to inspire a multinational enterprise, otherwise focused on the management

of landing fees and effluents, to underwrite a venture invested with such elevated artistic ambitions. Nevertheless, as the man from the airport company put it to me over the telephone, with a lyricism as vague as it was beguiling, there were still many aspects of the world that perhaps only writers could be counted on to find the right words to express. A glossy marketing brochure, while in certain contexts a supremely effective instrument of communication, might not always convey the authenticity achievable by a single authorial voice – or, as my friend suggested with greater concision, could more easily be dismissed as 'bullshit'.

2 Though the worlds of commerce and art have frequently been unhappy bedfellows, each viewing the other with a mixture of paranoia and contempt, I felt it would be churlish of me to decline to investigate my caller's offer simply because his company administered airside food courts and hosted technologies likely to be involved in raising the planet's median air temperature. There were undoubtedly some skeletons in the airport company's closet, arising from its intermittent desire to pour cement over age-old villages and its skill in encouraging us to circumnavigate the globe on unnecessary journeys, laden with bags of Johnnie Walker and toy bears dressed up as guards of the British monarchy.

But with my own closet not entirely skeleton-free, I was in no position to judge. I understood that money accumulated on the battlefield or in the marketplace could fairly be redirected towards higher aesthetic ends. I thought of impatient ancient Greek statesmen who had once spent their war spoils building temples to Athena and ruthless Renaissance noblemen who had blithely commissioned delicate frescoes in honour of spring.

Besides, and more prosaically, technological changes seemed to be drawing a curtain on a long and blessed interlude in which writers had been able to survive by selling their works to a wider public, threatening a renewed condition of anxious dependence on the largesse of individual sponsors. Contemplating what it might mean to be employed by an airport, I looked with plaintive

optimism to the example of the seventeenth-century philosopher Thomas Hobbes, who had thought nothing of writing his books while in the pay of the Earls of Devonshire, routinely placing florid declarations to them in his treatises and even accepting their gift of a small bedroom next to the vestibule of their home in Derbyshire, Hardwick Hall. 'I humbly offer my book to your Lordship,' England's subtlest political theorist had written to the swaggering William Devonshire on presenting him with *De Cive* in 1642. 'May God of Heaven crown you with many days in your earthly station, and many more in heavenly Jerusalem.'

In contrast, my own patron, Colin Matthews, the chief executive of BAA, the owner of Heathrow, was the most undemanding of employers. He made no requests whatever of me, not for a dedication, or even a modest reference to his prospects in the next world. His staff went so far as to give me explicit permission to be rude about the airport's activities. In such lack of constraints, I felt myself to be benefiting from a tradition wherein the wealthy merchant enters into a relationship with an artist fully prepared for him to behave like an outlaw; he does not expect good manners, he knows and is half delighted by the idea that the favoured baboon will smash his crockery. In such tolerance lies the ultimate proof of his power.

3 In any event, my new employer was legitimately proud of his terminal and understandably keen to find ways to sing of its beauty. The undulating glass and steel structure was the largest building in the land, forty metres tall and 400 long, the size of four football pitches, and yet the whole conveyed a sense of continuous lightness and ease, like an intelligent mind engaging effortlessly with complexity. The blinking of its ruby lights could be seen at dusk from Windsor Castle, the terminal's forms giving shape to the promises of modernity.

Standing before costly objects of technological beauty, we may be tempted to reject the possibility of awe, for fear that we could grow stupid through admiration. We may feel at risk of becoming overimpressed by architecture and engineering, of being dumbstruck by the Bombardier trains that progress

driverlessly between satellites or by the General Electric GE90 engines that hang lightly off the composite wings of a Boeing 777 bound for Seoul.

And yet to refuse to be awed at all might in the end be merely another kind of foolishness. In a world full of chaos and irregularity, the terminal seemed a worthy and intriguing refuge of elegance and logic. It was the imaginative centre of contemporary culture. Had one been asked to take a Martian to visit a single place that neatly captures the gamut of themes running through our civilisation – from our faith in technology to our destruction of nature, from our interconnectedness to our romanticising of travel – then it would have to be to the departures and arrivals halls that one would head. I ran out of reasons not to accept the airport's unusual offer to spend a little more time on its premises.

I   Approach

II   Departures

I arrived at the airport on a train from central London early on a Sunday evening, a small roller case in hand and no further destination for the week. I had been billeted at the Terminal 5 outpost of the Sofitel hotel chain, which, while not directly under the ownership of the airport, was situated only a few metres away from it, umbilically connected to the mothership by a sequence of covered walkways and a common architectural language featuring the repeated use of glazed surfaces, giant potted vegetation and grey tiling.

The hotel boasted 605 rooms that faced one another across an internal atrium, but it soon became evident that the true soul of the enterprise lay not so much in hostelry as in the management of a continuous run of conferences and congresses, held in forty-five meeting rooms, each one named after a different part of the world, and well equipped with data points and LAN facilities. At the end of this August Sunday, Avis Europe was in the Dubai Room and Liftex, the association of the British lift industry, in the Tokyo Hall. But the largest gathering was in the Athens Theatre, where delegates were winding up a meeting about valve sizes chaired by the International Organization for Standardization (or ISO), a body committed to eradicating incompatibilities between varieties of industrial equipment. So long as the Libyan government honoured its agreements, thanks to twenty years of work by the ISO, one would soon be able to travel across North Africa, from Agadir to El Gouna, without recourse to an adapter plug.

2 I had been assigned a room at the top western corner of the building, from which I could see the side of the terminal and a sequence of red and white lights that marked the end of the northern runway. Every minute, despite the best attempts of the glazing contractors, I heard the roar of an ascending jet, as hundreds of passengers, some perhaps holding their partners' hands, others sanguinely scanning *The Economist*, submitted themselves to a calculated defiance of our species' land-based origins. Behind each successful flight lay the coordinated efforts of hundreds of souls, from the manufacturers of airline amenity kits to the Honeywell engineers responsible for installing windshear-detection radars and collision-avoidance systems.

The hotel room appeared to have taken its design cues from the business-class cabin – though it was hard to say for sure which had inspired which, whether the room was skilfully endeavouring to look like a cabin, or the cabin a room – or whether they simply both shared in an unconscious spirit of their age, of the kind that had once ensured continuity between the lace trim on mid-eighteenth-century evening dresses and the iron detailing on the façades of Georgian town houses. The space held out the promise that its occupant might summon up a film on the adjustable screen, fall asleep to the drone of the air-conditioning unit and wake up on the final descent to Chek Lap Kok.

My employer had ordered me to remain within the larger perimeter of the airport for the duration of my seven-day stay and had accordingly provided me with a selection of vouchers from the terminal's restaurants as well as authorisation to order two evening meals from the hotel.

There can be few literary works in any language as poetic as a room-service menu.

> The autumn blast
> Blows along the stones
> On Mount Asama

Even these lines by Matsuo Bashō, who brought the haiku form to its mature perfection in the Edo era in Japan, seemed flat and unevocative next to the verse composed by the anonymous master at work somewhere within the Sofitel's catering operation:

> Delicate field greens with sun-dried cranberries,
> Poached pears, Gorgonzola cheese
> And candied walnuts in a Zinfandel vinaigrette

I reflected on the difficulty faced by the kitchen of correctly interpreting the likelihood of selling some of the remoter items of the menu: how many out of the guests in the lift industry, for example, might be tempted by the 'Atlantic snapper, enhanced with lemon pepper seasoning atop a chunky mango relish', or by the always mysterious and somewhat melancholy-sounding 'Chef's soup of the day'. But perhaps, in the end, there was no particular science to the calibration of alimentary supplies, for it is rare to spend an evening in a hotel and order anything other than a club sandwich, which even Bashō, at the peak of his powers, would have struggled to describe as convincingly as the menu's scribe:

Warm grilled chicken slices,
Smoked bacon, crisp lettuce,
And a warm ciabatta roll on a bed of sea-salted fries

There was a knock at the door only twenty minutes after I had dialled nine and put in my order. It is a strange moment when two adult men meet each other, one naked save for a complimentary dressing gown, the other (newly arrived in England from the small Estonian town of Rakvere and sharing a room with four others in nearby Hillingdon) sporting a black and white uniform, with an apron and a name badge. It is difficult to think of the ritual as entirely unremarkable, to say in a casually impatient voice, 'By the television, please,' while pretending to rearrange papers – though this capacity can be counted upon to evolve with more frequent attendance at global conferences.

I had dinner with Chloe Cho, formerly with Channel NewsAsia but now working for CNBC in Singapore. She updated me on the regional markets and Samsung's quarterly forecast, but her sustained focus was on commodities. I wondered what Chloe's outside interests might be. She was like a sister of the Carmelite Order, behind whose austere headdress and concentrated expression one could just guess at occasional moments of doubt, rendered all the more intriguing by their emphatic denial. On a ticker tape running across the bottom of the screen, I spotted the share price of my employer, pointed on a downward trajectory.

After dinner, it was still warm and not yet quite dark outside. I would have liked to take a walk around one of the few fields that remained of the farmland on which the airport had been built some six decades before, but it seemed at once perilous and impossible to leave the building, so I decided to do a few circuits around the hotel corridors instead. Feeling disoriented and queasy, as if I were on a cruise ship in a swell, I repeatedly had to steady myself against the synthetic walls. Along my route, I passed dozens of room-service trays much like my own, each one furtively pushed into the hallway and nearly all (once their stainless-steel covers were lifted) providing evidence of orgiastic episodes of consumption. Ketchup smeared across slices of toast and fried eggs dipped in vinaigrette spoke of the breaking of taboos just like the sexual ones more often assumed to be breached during solitary residence in hotel rooms.

I fell asleep at eleven, but woke up again abruptly just past three. The prehistoric part of the mind, trained to listen for and interpret every shriek in the trees, was still doing its work, latching on to the slamming of doors and the flushing of toilets in unknown precincts of the building. The hotel and terminal seemed like a giant machine poised in standby mode, emitting an uncanny hum from a phalanx of slowly rotating exhaust fans. I thought of the hotel's spa, its hot tubs perhaps still bubbling in the darkness. The sky was a chemical orange colour, observing the final hours of the fragile curfew it had been keeping ever since it had swallowed up the last of the previous evening's Asia-bound flights. Jutting from the side of the terminal was the disembodied tail of a British Airways A321, anticipating another imminent odyssey in the merciless cold of the lower stratosphere.

3   In the end it was a 5.30 a.m. arrival (BA flight from Hong Kong) that called a halt to my perturbed night. I showered, ate a fruit bar purchased from a dispensing machine in the car park and wandered over to an observation area next to the terminal. In the cloudless dawn, a sequence of planes, each visible as a single diamond, were lined up at different heights, like pupils in a school photo, on their final approach to the northern runway. Their wings unfolded themselves into elaborate and unlikely

arrangements of irregularly sized steel-grey panels. Having avoided the earth for so long, wheels that had last touched ground in San Francisco or Mumbai hesitated and slowed almost to a standstill as they arched and prepared to greet the rubber-stained English tarmac with a burst of smoke that made manifest their planes' speed and weight.

With the aggressive whistling of their engines, the airborne visitors appeared to be rebuking this domestic English morning for its somnolence, like a delivery person unable to resist pressing a little too insistently and vengefully on the doorbell of a still-slumbering household. All around them, the M4 corridor was waking up reluctantly. Kettles were being switched on in Reading, shirts being ironed in Slough and children unfurling themselves beneath their Thomas the Tank Engine duvets in Staines.

Yet for the passengers in the 747 now nearing the airfield, the day was already well advanced. Many would have awakened several hours before to see their plane crossing over Thurso at the northernmost tip of Scotland, nearly the end of the earth to those in London's suburbs, but their destination's very doorstep for travellers after a long night's journey over the Canadian icelands and a moonlit North Pole. Breakfast would have kept time with the airliner's progress down the spine of the kingdom: a struggle with a small box of cornflakes over Edinburgh, an omelette studded with red peppers and mushrooms near Newcastle, a stab at a peculiar-looking fruit yoghurt over the unknowing Yorkshire Dales.

For British Airways planes, the approach to Terminal 5 was a return to their home base, equivalent to the final run up the Plymouth Sound for their eighteenth-century naval predecessors. Having long been guests on foreign aprons, allotted awkward and remote slots at O'Hare or LAX, the odd ones out amid immodestly long rows of United and Delta aircraft, they now took their turn at having the superiority of numbers, lining up in perfect symmetry along the back of Satellite B.

Sibling 747s that had only recently been separated out across the world were here parked wing tip to wing tip, Johannesburg next to Delhi, Sydney next to Phoenix. Repetition lent their fuselage designs a new beauty: the eye could follow a series of identical motifs down a fifteen-strong line of dolphin-like bodies, the resulting aesthetic effect only enhanced by the knowledge that each plane had cost some $250 million, and that what lay before one was therefore a symbol not just of the modern era's daunting technical intelligence but also of its prodigious and inconceivable wealth.

As every plane took up its position at its assigned gate, a choreographed dance began. A passenger walkway rolled forward and closed its rubber mouth in a hesitant kiss over the front left-hand door. A member of the ground staff tapped at the window, a colleague inside released the airlock and the two airline personnel exchanged the sort of casual greeting one might have expected between office workers returning to adjacent desks after lunch, rather than the encomium that would more fittingly have marked the end of an 11,000-kilometre journey from the other side of the globe. Then again, the welcome may be no more effusive a hundred years hence, when, at the close of a nine-month voyage, against the eerie blood-red midday light bathing a spaceport in Mars's Cydonian hills, a fellow human knocks at the gold-tinted window of our just-docked craft.

Cargo handlers opened the holds to unload crates filled with the chilled flanks of Argentine cattle and the crenellated forms of crustaceans that had, just the day before, been marching heedlessly across Nantucket Sound. In only a few hours, the plane would be sent up into the sky once more. Fuel hoses were attached to its wings and the tanks replenished with Jet A-1 that would steadily be burned over the African savannah. In the already vacant front cabins, where it might cost the equivalent of a small car to spend the night reclining in an armchair, cleaners scrambled to pick up the financial weeklies, half-eaten chocolates and distorted foam earplugs left behind by the flight's complement of plutocrats and actors. Passengers disembarked for whom this ordinary English morning would have a supernatural tinge.

4   Meanwhile, at the drop-off point in front of the terminal, cars were pulling up in increasing numbers, rusty minicabs with tensely negotiated fares alongside muscular limousines from whose armoured doors men emerged crossly and swiftly into the executive channels.

   Some of the trips starting here had been decided upon only in the previous few days, booked in response to a swiftly developing situation in the Munich or Milan office; others were the fruit of three years' painful anticipation of a return to a village in northern Kashmir, with six dark-green suitcases filled with gifts for young relatives never previously met.

The wealthy tended to carry the least luggage, for their rank and itineraries led them to subscribe to the much-published axiom that one can now buy anything anywhere. But they had perhaps never visited a television retailer in Accra or they might have looked more favourably upon a Ghanaian family's decision to import a Samsung PS50, a high-definition plasma machine the weight and size of a laden coffin. It had been acquired the day before at a branch of Comet in Harlow and was eagerly awaited in the Kissehman quarter of Accra, where its existence would stand as evidence of the extraordinary status of its importer, a thirty-eight-year-old dispatch driver from Epping.

Entry into the vast space of the departures hall heralded the opportunity, characteristic in the transport nodes of the modern world, to observe people with discretion, to forget oneself in a sea of otherness and to let the imagination loose on the limitless supply of fragmentary stories provided by the eye and ear. The mighty steel bracing of the airport's ceiling recalled the scaffolding of the great nineteenth-century railway stations, and evoked the sense of awe – suggested in paintings such as Monet's *Gare Saint-Lazare* – that must have been experienced by the first crowds to step inside these light-filled, iron-limbed halls pullulating with strangers, buildings that enabled a person to sense viscerally, rather than just grasp intellectually, the vastness and diversity of humanity.

The roof of the building weighed 18,000 tonnes, but the steel columns supporting it hardly suggested the pressures they were under. They were endowed with a subcategory of beauty we might refer to as elegance, present whenever architecture has the modesty not to draw attention to the difficulties it has surmounted. On top of their tapered necks, the columns balanced the 400-metre roof as if they were holding up a canopy made of linen, offering a metaphor for how we too might like to stand in relation to our burdens.

Most passengers were bound for a bank of automatic check-in machines in the centre of the hall. These represented an epochal shift away from the human hand and towards the robot, a transition as significant in the context of airline logistics as that from the washboard to the washing machine had once been in the domestic sphere. However, few users seemed capable of producing the precise line-up of cards and codes demanded by the computers, which responded to the slightest infraction with sudden and intemperate error messages – making one long for a return of the surliest of humans, from whom there always remains at least a theoretical possibility of understanding and forgiveness.

Nowhere was the airport's charm more concentrated than on the screens placed at intervals across the terminal which announced, in deliberately workmanlike fonts, the itineraries of aircraft about to take to the skies. These screens implied a feeling of infinite and immediate possibility: they suggested the ease with which we might impulsively approach a ticket desk and, within a few hours, embark for a country where the call to prayer rang out over shuttered whitewashed houses, where we understood nothing of the language and where no one knew our identities. The lack of detail about the destinations served only to stir unfocused images of nostalgia and longing: Tel Aviv, Tripoli, St Petersburg, Miami, Muscat via Abu Dhabi, Algiers, Grand Cayman via Nassau... all of these promises of alternative lives, to which we might appeal at moments of claustrophobia and stagnation.

5 A few zones of the check-in area remained dedicated to traditionally staffed desks, where passengers were from the start assured of interaction with a living being. The quality of this

interaction was the responsibility of Diane Neville, who had worked for British Airways since leaving school fifteen years before and now oversaw a staff of some two hundred who dispensed boarding cards and affixed luggage labels.

It was never far from Diane's thoughts how vulnerable her airline was to its employees' bad moods. On reaching home, a passenger would remember nothing of the plane that had not crashed or the suitcase that had arrived within minutes of the carousel's starting if, upon politely asking for a window seat, she had been brusquely admonished to be happy with whatever she was assigned – this retort stemming from a sense on the part of a member of the check-in team (perhaps discouraged by a bad head cold or a disappointing evening at a nightclub) of the humiliating and unjust nature of existence.

In the earliest days of industry, it had been an easy enough matter to motivate a workforce, requiring only a single and basic tool: the whip. Workers could be struck hard and with impunity to encourage them to quarry stones or pull on their oars with greater enthusiasm. But the rules had had to be revised with the development of jobs – by the early twenty-first century comprising the dominant sector of the market – that could be successfully performed only if their protagonists were to a significant degree satisfied rather than resentfully obedient. Once it became evident that someone who was expected to wheel elderly passengers around a terminal, for example, or to serve meals at high altitudes could not profitably be sullen or furious, the mental well-being of employees began to be a supreme object of commercial concern.

Out of such requirements had been born the art of management, a set of practices designed to coax rather than simply extort commitment out of workers, and which, at British Airways, had inspired the use of regular motivational training seminars, gym access and free cafeterias in order to achieve that most calculated, unsentimental and fragile of goals: a friendly manner.

But however skilfully designed its incentive structure, the airline could in the end do very little to guarantee that its staff would actually add to their dealings with customers that almost imperceptible measure of goodwill which elevates service from mere efficiency to tangible warmth. Though one can inculcate competence, it is impossible to legislate for humanity. In other words, the airline's survival depended upon qualities that the

7 Not far from the incautiously hopeful man, a pair of lovers were parting. She must have been twenty-three, he a few years older. There was a copy of Haruki Murakami's *Norwegian Wood* in her bag. They both wore oversize sunglasses and had come of age in the period between SARS and swine flu. It was the intensity of their kiss that first attracted my attention, but what had seemed like passion from afar was revealed at closer range to be an unusual degree of devastation. She was shaking with sorrowful disbelief as he cradled her in his arms and stroked her wavy black hair, in which a clip shaped like a tulip had been fastened. Again and again, they looked into each other's eyes and every time, as though made newly aware of the catastrophe about to befall them, they would begin weeping once more.

Passers-by evinced sympathy. It helped that the woman was extraordinarily beautiful. I missed her already. Her beauty would have been an important part of her identity from at least the age of twelve and, in its honour, she would occasionally pause and briefly consider the effect of her condition on her audience before returning to her lover's chest, damp with her tears.

We might have been ready to offer sympathy, but in actuality there were stronger reasons to want to congratulate her for

having such a powerful motive to feel sad. We should have envied her for having located someone without whom she so firmly felt she could not survive, beyond the gate let alone in a bare student bedroom in a suburb of Rio. If she had been able to view her situation from a sufficient distance, she might have been able to recognise this as one of the high points in her life.

There seemed no end to the ritual. The pair would come close to the security zone, then break down again and retreat for another walk around the terminal. At one point, they went down to the arrivals hall and for a moment it looked as if they might go outside and join the queue at the taxi rank, but they were only buying a packet of dried mango slices from Marks and Spencer, which they fed to each other with pastoral innocence. Then quite suddenly, in the middle of an embrace by the Travelex desk, the beauty glanced down at her watch and, with all the self-control of Odysseus denying the Sirens, ran away from her tormentor down a corridor and into the security zone.

My photographer and I divided forces. I followed her airside and watched her remain stoic until she reached the concourse, only to founder again at the window of Kurt Geiger. I finally lost her in a crowd of French exchange students near Sunglass Hut. For his part, Richard pursued the man down to the train station, where the object of adoration boarded the express service for central London, claimed a seat and sat impassively staring out the window, betraying no sign of emotion save for an unusual juddering movement of his left leg.

8 For many passengers, the terminal was the starting point of short-haul business trips around Europe. They might have announced to their colleagues a few weeks before that they would be missing a few days in the office to fly to Rome, studiously feigning weariness at the prospect of making a journey to the wellspring of European culture – albeit to its frayed edges in a business park near Fiumicino airport.

They would think of these colleagues as they crossed over the Matterhorn, its peak snow-capped even in summer. Just as breakfast was being served in the cabin, their co-workers would be coming into the office – Megan with her carefully prepared lunch, Geoff with his varied ring tones, Simi with her permanent frown – and all the while the travellers would be witnessing below them the byproducts of the titanic energies released by the collision of the Eurasian and African continental plates during the late Mesozoic era.

What a relief it would be for the travellers not to have time to see anything at all of Rome's history or art. And yet how much they would notice nevertheless: the fascinating roadside advertisements for fruit juice on the way from the airport, the unusually delicate shoes worn by Italian men, the odd inflections in their hosts' broken English. What interesting new thoughts would occur to them in the Novotel, what inappropriate films they would watch late into the night and how heartily they would agree, upon their return, with the truism that the best way to see a foreign country is to go and work there.

9   A full 70 per cent of the airport's departing passengers were off on trips for pleasure. It was easy to spot them at this time of year, in their shorts and hats. David was a thirty-eight-year-old shipping broker, and his wife, Louise, a thirty-five-year-old full-time mother and ex-television producer. They lived in Barnes with their two children, Ben, aged three, and Millie, aged five. I found them towards the back of a check-in line for a four-hour flight to Athens. Their final destination was a villa with a pool at the Katafigi Bay resort, a fifty-minute drive away from the Greek capital in a Europcar Category C vehicle.

It would be difficult to overestimate how much time David had spent thinking about his holiday since he had first booked it, the previous January. He had checked the weather reports online every day. He had placed the link to the Dimitra Residence in his Favourites folder and regularly navigated to it, bringing up images of the limestone master bathroom and of the house at dusk, lit up against the rocky Mediterranean slopes. He had pictured himself playing with the children in the palm-lined garden and eating grilled fish and olives with Louise on the terrace.

But although David had reflected at length on his stay in the Peloponnese, there were still many things that managed to surprise him at Terminal 5. He had omitted to recall the existence of the check-in line or to think of just how many people can be fitted into an Airbus A320. He had not focused on how long four hours can seem nor had he considered the improbability of all the members of a family achieving physical and psychological satisfaction at approximately the same time. He had not remembered how hurtful he always found it when Ben made it clear that he disproportionately favoured his mother or how he himself invariably responded to such rejections by becoming unproductively strict, which in turn upset his wife, who liked

to voice her opinion that Ben's reticence was due primarily to the lack of paternal contact he had had since his father's promotion. David's work was a continuous flash-point in the couple's relationship and had in fact precipitated an argument only the night before, during which David had described Louise as ungrateful for failing to appreciate and honour the necessary connection between his absences and their affluence.

Had the plane on which they were to fly to Athens burst into flames shortly after take-off and begun plunging towards the Staines reservoir, David would have clasped the members of his family tightly to him and told them with wholehearted sincerity that he loved them unreservedly – but right now, he could not look a single one of them in the eye.

It seems that most of us could benefit from a brush with a near-fatal disaster to help us to recognise the important things that we are too defeated or embittered to recognise from day to day.

As David lifted a suitcase on to the conveyor belt, he came to an unexpected and troubling realisation: that he was bringing *himself* with him on his holiday. Whatever the qualities of the Dimitra Residence, they were going to be critically undermined by the fact that *he* would be in the villa as well. He had booked the trip in the expectation of being able to enjoy his children, his wife, the Mediterranean, some spanakopita and the Attic skies, but it was evident that he would be forced to apprehend all of these through the distorting filter of his own being, with its debilitating levels of fear, anxiety and wayward desire.

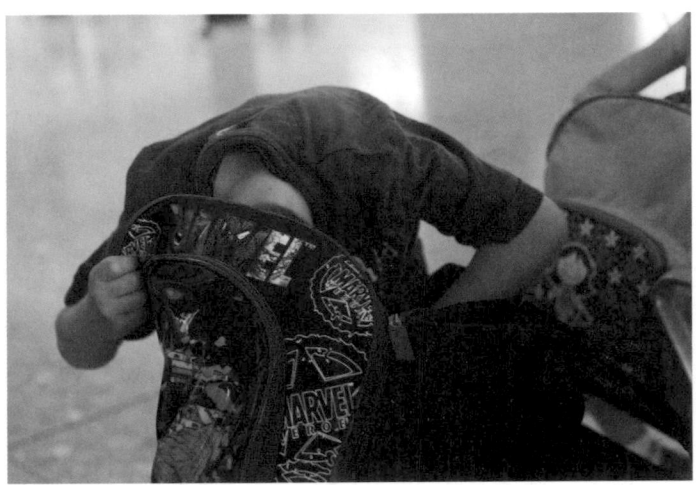

There was, of course, no official recourse available to him, whether for assistance or complaint. British Airways did, it was true, maintain a desk manned by some unusually personable employees and adorned with the message: 'We are here to help'. But the staff shied away from existential issues, seeming to restrict their insights to matters relating to the transit time to adjacent satellites and the location of the nearest toilets.

Yet it was more than a little disingenuous for the airline to deny all knowledge of, and responsibility for, the metaphysical well-being of its customers. Like its many competitors, British Airways, with its fifty-five Boeing 747s and its thirty-seven Airbus A320s, existed in large part to encourage and enable people to go and sit in deckchairs and take up (and usually fail at) the momentous challenge of being content for a few days. The tense atmosphere now prevailing within David's family was a reminder of the rigid, unforgiving logic to which human moods are subject, and which we ignore at our peril when we see a picture of a beautiful house in a foreign country and imagine that happiness must inevitably accompany such magnificence. Our capacity to derive pleasure from aesthetic or material goods seems critically dependent on our first satisfying a more important range of emotional and psychological needs, among them those for understanding, compassion and respect. We cannot enjoy palm trees and azure pools if a relationship to which we are committed has abruptly revealed itself to be suffused with incomprehension and resentment.

There is a painful contrast between the enormous objective projects that we set in train, at incalculable financial and environmental cost – the construction of terminals, of runways and of wide-bodied aircraft – and the subjective psychological knots that undermine their use. How quickly all the advantages of technological civilisation are wiped out by a domestic squabble. At the beginning of human history, as we struggled to light fires and to chisel fallen trees into rudimentary canoes, who could have predicted that long after we had managed to send men to the moon and aeroplanes to Australasia, we would still have such trouble knowing how to tolerate ourselves, forgive our loved ones and apologise for our tantrums?

10 My employer had made good on the promise of a proper desk. It turned out to be an ideal spot in which to do some work, for it rendered the idea of writing so unlikely as to make it possible again. Objectively good places to work rarely end up being so: in their faultlessness, quiet and well-equipped studies have a habit of rendering the fear of failure overwhelming. Original thoughts are like shy animals. We sometimes have to look the other way – towards a busy street or terminal – before they run out of their burrows.

The setting was certainly rich in distractions. Every few minutes, a voice (usually belonging to either Margaret or her colleague Juliet, speaking from a small room on the floor below) would make an announcement attempting, for example, to reunite a Mrs Barker, recently arrived from Frankfurt, with a stray piece of her hand luggage or reminding Mr Bashir of the pressing need for him to board his flight to Nairobi.

As far as most passengers were concerned, I was an airline employee and therefore a potentially useful source of information on where to find the customs desk or the cash machine. However, those who took the trouble to look at my name badge soon came to regard my desk as a confessional.

One man came to tell me that he was embarking on what he wryly termed the holiday of a lifetime to Bali with his wife, who was just months away from succumbing to incurable brain cancer. She rested nearby, in a specially constructed wheelchair laden with complicated breathing apparatus. She was forty-nine years old and had been entirely healthy until the previous April, when she had gone to work on a Monday morning complaining of a slight headache. Another man explained that he had been visiting his wife and children in London, but that he had a second family in Los Angeles who knew nothing about the first. He had five children in all, and two mothers-in-law, yet his face bore none of the strains of his situation.

Each new day brought such a density of stories that my sense of time was stretched. It seemed like weeks, though it was in fact just a couple of days, since I had met Ana D'Almeida and Sidonio Silva, both from Angola. Ana was headed for Houston, where she was studying business, and Sidonio for Aberdeen, where he was completing a PhD in mechanical engineering. We spent an hour together, during which they spoke in idealistic and melancholy ways of the state of their country. Two days

later, Heathrow held no memories of them, but I felt their absence still.

There were some more permanent fixtures in the terminal. My closest associate was Ana-Marie, who cleaned the section of the check-in area where my desk had been set up. She said she was eager to be included in my book and stopped by several times to chat with me about the possibility. But when I assured her that I would write something about her, a troubled look came over her face and she insisted that I would have to disguise her real name and features. The truth would disappoint too many of her friends and relatives back in Transylvania, she said, for as a young woman she had been the leading student in her conservatoire and since then was widely thought to have achieved renown abroad as a classical singer.

The presence of a writer occasionally raised expectations that something dramatic might be on the verge of occurring, the sort of thing one could read about in a novel. My explanation that I was merely looking around, and required nothing more extraordinary of the airport than that it continue to operate much as it did every other day of the year, was sometimes greeted with disappointment. But the writer's desk was at heart an open invitation to users of the terminal to begin studying

their setting with a bit more imagination and attention, to give weight to the feelings that airports provoke, but which we are seldom able to sort through or elaborate upon in the anxiety of making our way to the gate.

My notebooks grew thick with anecdotes of loss, desire and expectation, snapshots of travellers' souls on their way to the skies – though it was hard to dismiss a worry about what a modest and static thing a book would always be next to the chaotic, living entity that was a terminal.

11 At moments when I could not make headway with my writing, I would go and chat to Dudley Masters, who was based on the floor below me and had spent thirty years cleaning shoes at the airport. His day began at 8.30 a.m. and, around sixty pairs later, finished at 9.00 p.m.

I admired the optimism with which Dudley confronted every new pair of shoes that paused at his station. Whatever their condition, he imagined the best for them, remedying their abuses with an armoury of brushes, waxes, creams and spray cleaners. He knew it was not evil that led people to go for eight months without applying even an all-purpose clear cream polish. He was like a kindly dentist who, on bringing down the ceiling-mounted halogen lamp and asking new patients to open their mouths ('Let's have a look in here, shall we?'), remains aware of how complicated lives can become and so how easily people may give up flossing their teeth while they try to save their companies or minister to a dying parent.

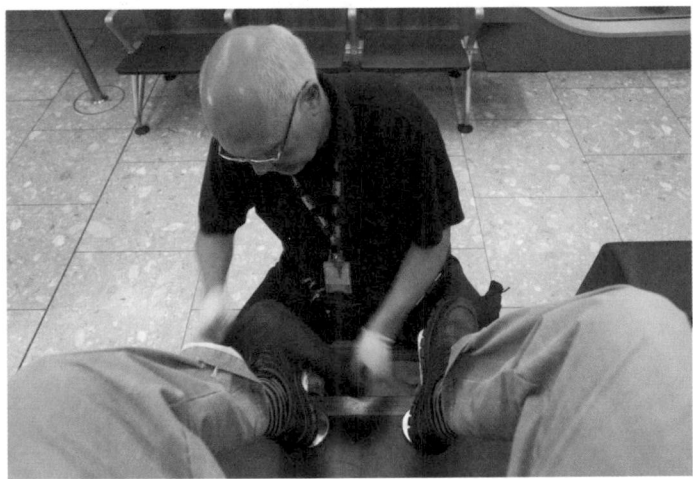

Though he was being paid to shine shoes, he knew that his real mission was psychological. He understood that people rarely have their shoes cleaned at random: they do so when they want to draw a line under the past, when they hope that an outer transformation may be a spur to an inner one. With no ill will, nor any desire to taunt me, he would daily assure me that if he ever got around to putting his experiences down on paper, his would be the most fascinating book about an airport that anyone had ever read.

12 Just past Dudley's workstation, off a corridor leading to the security zone, there was a multi-faith room, a cream-coloured space holding an ill-matched assortment of furniture and a bookshelf of the sacred texts.

I watched a family from southern India coming to pay their respects to Ganesh, the Hindu god in charge of the fortunes of travellers, before going on to board the 1.00 p.m. BA035 flight to Chennai. The deity was presented with some cupcakes and a rose-scented candle, which airport regulations prevented the family from actually lighting.

In the old days, when aircraft routinely fell out of the sky because large and obvious components failed – the fuel pumps gave out or the engines exploded – it felt sensible to cast aside the claims of organised religions in favour of a trust in science. Rather than praying, the urgent task was to study the root causes of malfunctions and stamp out error through reason. But as aviation has become ever more subject to scrutiny, as every part has been hedged by backup systems, so, too, have the reasons for becoming superstitious paradoxically increased.

The sheer remoteness of a catastrophic event occurring invites us to forgo scientific assurances in favour of a more humble stance towards the dangers which our feeble minds struggle to contain. While never going so far as to ignore maintenance schedules, we may nevertheless judge it far from unreasonable to take a few moments before a journey to fall to our knees and pray to the mysterious forces of fate to which all aircraft remain subject and which we might as well call Isis, God, Fortuna or Ganesh – before going on to buy cigarettes and Chanel No.5 in the World Duty Free emporium on the other side of security.

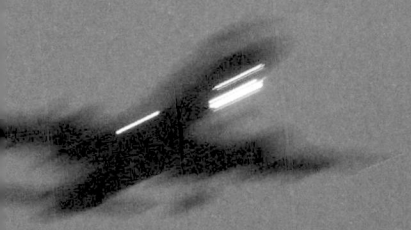

III Airside

1 The security line was impressive as always, numbering at least a hundred people reconciled, though with varying degrees of acceptance, to the idea of not doing very much else with the next twenty minutes of their lives.

    The station furthest to the left was staffed by Jim at the scanner, Nina at the manual bag check and Balanchandra at the metal detector. Each had submitted to an arduous year-long course, the essential purpose of which was to train them to look at every human being as though he or she might want to blow up an aircraft – a thoroughgoing reversal of our more customary impulse to find common ground with new acquaintances. The team had been taught to overcome all prejudices as to what an enemy might look like: it could well be the six-year-old girl holding a carton of apple juice and her mother's hand or the frail grandmother flying to Zurich for a funeral. Suspects, guilty until proved innocent, would therefore need to be told in no uncertain terms to step aside from their belongings and stand straight up against the wall.

    Like thriller writers, the security staff were paid to imagine life as a little more eventful than it customarily manages to be. I felt sympathy for them in their need to remain alert at every moment of their careers, perpetually poised to react to the most remote of possibilities, of the sort that occurred globally in their line of work perhaps only once in a decade, and even then probably in Larnaca or Baku. They were like members of an evangelical sect living in a country devoid of biblical precedents – Belgium, say, or New Zealand – whose beliefs had inspired a daily expectation of a local return of the Messiah, a prospect not to be discounted even at 3.00 p.m. on a Wednesday in suburban Liège. How enviously the staff must have considered the lot of ordinary policemen and women, who, despite their often unsociable hours and wearying foot patrols, could at least look forward to having regular encounters with exactly the sorts of characters whom they had been trained to deal with.

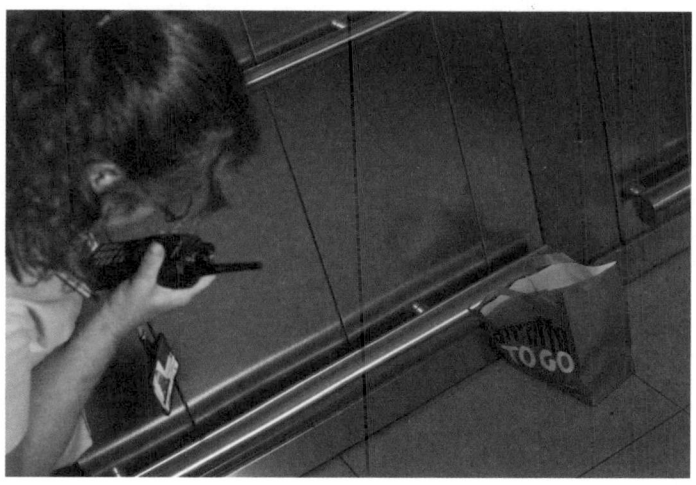

I felt additional sympathy for the staff as a result of the limited curiosity they were permitted to bring to bear on the targets of their searches. Despite having free rein to look inside any passenger's make-up bag, diary or photo album, they were allowed to investigate only evidence pointing to the presence of explosive devices or murder weapons. There was therefore no sanction for them to ask for whom a neatly wrapped package of underwear was intended, nor any official recognition of how tempting it might occasionally seem to stroke the back pockets of a pair of low-slung jeans without any desire to discover a semi-automatic pistol.

So great was the pressure imposed on the team by the need for vigilance that they were granted more frequent tea breaks than other employees. Every hour they would repair to a room fitted out with dispensing machines, frayed armchairs and pictures of the world's most-wanted terrorists, a series of angry-looking, prophet-like figures with long beards and inscrutable eyes, apparently holed up in mountain caves and reluctant ever to venture into Terminal 5.

It was in this room that I spotted two women who looked as if they might be students enrolled in some sort of internship programme. When I smiled at them, hoping thereby to make them feel a bit more welcome, they came over to greet me and introduced themselves as the two most senior security officers in the building. In charge of training for the entire security staff at Terminal 5, Rachel and Simone regularly taught teams how to disarm terrorists and what positions to adopt in order to protect themselves in the event of a grenade being thrown. They also gave individual employees basic instruction in the use of semi-automatic weapons. Their close focus on anti-terrorism seemed to colour all aspects of their lives: in their spare time, they both read whatever literature they could find on the subject. Rachel was a specialist in the 1976 Entebbe operation, Simone a keen student of the Hindawi Affair, in which a Jordanian man, Nezar Hindawi, had given a Semtex-filled bag to his pregnant girlfriend and persuaded her to board an El Al plane for Tel Aviv. Though the plot had failed, Simone explained (unknowingly damning my naïve conclusions on the wisdom of bothering to search certain sorts of passengers), the incident had forever changed the way security personnel the world over would look at pregnant women, small children and kindly grandmothers.

If many passengers became anxious or angry upon being questioned or searched, it was because such investigations could easily begin to feel, if only on a subconscious level, like accusations, and might thereby slot into pre-existing proclivities towards a sense of guilt.

A long wait for a scanning machine can induce many of us to start asking ourselves if we have perhaps after all left home with an explosive device hidden in our case, or unwittingly submitted to a months-long terrorist training course. The psychoanalyst

Melanie Klein, in her *Envy and Gratitude* (1963), traced this latent sense of guilt back to an intrinsic part of human nature, originating in our Oedipal desire to murder our same-sex parent. So strong can the guilty feeling become in adulthood that it may provoke a compulsion to make false confessions to those in authority, or even to commit actual crimes as a means of gaining a measure of relief from an otherwise overwhelming impression of having done something wrong.

Safe passage through security did have one advantage, at least for those plagued (like the author) by a vague sense of their own culpability. A noiseless, unchecked progress through the detectors allowed one to advance into the rest of the terminal with a feeling akin to that one may experience on leaving church after confession or synagogue on the Day of Atonement, momentarily absolved and relieved of some of the burden of one's sins.

2 There was a good deal of shopping to be done on the other side of security, where more than one hundred separate retail outlets vied for the attention of travellers – a considerably greater number than were to be found in the average shopping centre. This statistic regularly caused critics to complain that Terminal 5 was more like a mall than an airport, though it was hard to determine what might be so wrong with this balance, what precise aspect of the building's essential aeronautical identity had been violated or even what specific pleasure passengers had been robbed of, given that we are inclined to visit malls even when they don't provide us with the additional pleasure of a gate to Johannesburg.

At the entrance to the main shopping zone was a currency-exchange desk. Although we are routinely informed that we live in a vast and diverse world, we may do little more than nod distractedly at this idea until the moment comes when we find ourselves at the back of a bureau de change lined with a hundred safe-deposit boxes, some containing neat sheaves of Uruguayan pesos, Turkmenistani manats and Malawian kwachas. The trading desks of the City of London might perform their transactions with incomparable electronic speed, but patient physical contact with thick bundles of notes offered a very different sort of immediacy: a living sense of the miscellany of the human species. These notes, in every colour and font, were decorated with images of strongmen, dictators, founding fathers, banana trees and leprechauns. Many were worn and creased from heavy use. They had helped to pay for camels in Yemen or saddles in Peru, been stashed in the wallets of elderly barbers in Nepal or under the pillows of schoolboys in Moldova. A fraying fifty-kina note from Papua New Guinea (bird of paradise on the back, Prime Minister Michael Somare on the front) hardly hinted at the sequence of transactions (from fruit to shoes, guns to toys) that had culminated in its arrival at Heathrow.

Across the way from the exchange desk was the terminal's largest bookshop. Seemingly in spite of the author's defensive predictions about the commercial future of books (perhaps linked to the unavailability of any of his titles at any airport outlet), sales here were soaring. One could buy two volumes and get a third for free, or pick up four and be eligible for a fizzy drink. The death of literature had been exaggerated. Whereas on dating websites, those who like books are usually bracketed into a single category, the broad selections on offer at WH Smith spoke to the diversity of individuals' motives for reading. If there was a conclusion to be drawn from the number of bloodstained covers, however, it was that there was a powerful desire, in a wide cross-section of airline passengers, to be terrified. High above the earth, they were looking to panic about being murdered, and thereby to forget their more mundane fears about the success of a conference in Salzburg or the challenges of having sex for the first time with a new partner in Antigua.

I had a chat with a manager named Manishankar, who had been working at the shop since the terminal first opened. I explained – with the excessive exposition of a man spending a lonely week at the airport – that I was looking for the sort of books in which a genial voice expresses emotions that the reader has long felt but never before really understood; those that convey the secret, everyday things that society at large prefers to leave unsaid; those that make one feel somehow less alone and strange.

Manishankar wondered if I might like a magazine instead. There was no shortage, including several with feature articles on how to look good after forty – advice of course predicated on the assumption that one's appearance had been pleasing at thirty-nine (the writer's age).

Nearby, another bookcase held an assortment of classic novels, which had been imaginatively arranged, not by author or title, but according to the country in which their narratives were set. Milan Kundera was being suggested as a guide to Prague, and Raymond Carver depended upon to reveal the hidden character of the small towns between Los Angeles and Santa Fe. Oscar Wilde once remarked that there had been less fog in London before James Whistler started to paint, and one wondered if the silence and sadness of isolated towns in the American West had not been similarly less apparent before Carver began to write.

Every skilful writer foregrounds notable aspects of experience, details that might otherwise be lost in the mass of data that continuously bathes our senses – and in so doing prompts us to find and savour these in the world around us. Works of literature could be seen, in this context, as immensely subtle instruments by means of which travellers setting out from Heathrow might be urged to pay more careful attention to such things as the conformity and corruption of Cologne society (Heinrich Böll), the quiet eroticism of provincial Italy (Italo Svevo) or the melancholy of Tokyo's subways (Kenzaburō Ōe).

3 It was only after several days of frequenting the shops that I started to understand what those who objected to the dominance of consumerism at the airport might have been complaining about. The issue seemed to centre on an incongruity between shopping and flying, connected in some sense to the desire to maintain dignity in the face of death.

Despite the many achievements of aeronautical engineers over the last few decades, the period before boarding an aircraft is still statistically more likely to be the prelude to a catastrophe than a quiet day in front of the television at home. It therefore tends to raise questions about how we might best spend the last moments before our disintegration, in what frame of mind we might wish to fall back down to earth – and the extent to which we would like to meet eternity surrounded by an array of duty-free bags.

Those who attacked the presence of the shops might in essence have been nudging us to prepare ourselves for the end.

At the Blink beauty bar, I felt anew the relevance of the traditional religious call to seriousness voiced in Bach's Cantata 106:

> *Bestelle dein Haus,*
> *Denn du wirst Sterben,*
> *Und nicht lebendig bleiben.*

> Set thy house in order,
> For thou shalt die,
> And not remain alive.

Despite its seeming mundanity, the ritual of flying remains indelibly linked, even in secular times, to the momentous themes of existence – and their refractions in the stories of the world's religions. We have heard about too many ascensions, too many voices from heaven, too many airborne angels and saints to ever be able to regard the business of flight from an entirely pedestrian perspective, as we might, say, the act of travelling by train. Notions of the divine, the eternal and the significant accompany us covertly on to our craft, haunting the reading aloud of the safety instructions, the weather announcements made by our captains and, most particularly, our lofty views of the gentle curvature of the earth.

4 It seemed appropriate that I should bump into two clergymen just outside a perfume outlet, which released the gentle, commingled smell of some eight thousand varieties of scent. The older of the pair, the Reverend Sturdy, wore a high-visibility jacket with the words 'Airport Priest' printed on the back. In his late sixties, he had a vast and archetypically ecclesiastical beard and gold-rimmed spectacles. The cadence of his speech was impressively slow and deliberate, like that of a scholar unable to ignore, even for a moment, the nuances behind every statement, and accustomed to living in environments where these could be investigated to their furthest conclusions without fear of inconveniencing or delaying others. His colleague, Albert Kahn, likewise garbed for high visibility – though his jacket, borrowed from another staff member, read merely 'Emergency Services' – was in his early twenties and on a work placement at Heathrow while completing theological studies at Durham University.

'What do people tend to come to you to ask?' I enquired of the Reverend Sturdy as we passed by an outlet belonging to that perplexingly indefinable clothing brand Reiss. There was a long pause, during which a disembodied voice reminded us once more never to leave our luggage unattended.

'They come to me when they are lost,' the Reverend replied at last, emphasising the final word so that it seemed to reflect the spiritual confusion of mankind, a hapless race of beings described by St Augustine as 'pilgrims in the City of Earth until they can join the City of God'.

'Yes, but what might they be feeling lost *about?*'

'Oh,' said the Reverend with a sigh, 'they are almost always looking for the toilets.'

Because it seemed a pity to end our discussion of metaphysical matters on such a note, I asked the two men to tell me how a traveller might most productively spend his or her last minutes before boarding and take-off. The Reverend was adamant: the task, he said, was to turn one's thoughts intently to God.

'But what if one can't believe in him?' I pursued.

The Reverend fell silent and looked away, as though this were not a polite question to ask of a priest. Happily, his colleague, weaned on a more liberal theology, delivered an equally succinct but more inclusive reply, to which my thoughts often returned in the days to come as I watched planes taxiing out to the runways: 'The thought of death should usher us towards whatever happens to matter most to us; it should lend us the courage to pursue the way of life we value in our hearts.'

5

Just beyond the security area was a suite, named after an ill-fated supersonic jet and reserved for the use of first-class passengers. The advantages of wealth can sometimes be hard to see: expensive cars and wines, clothes and meals are nowadays rarely proportionately superior to their cheaper counterparts, due to the sophistication of modern processes of design and mass production. But in this sense, British Airways' Concorde Room was an anomaly. It was humblingly and thought-provokingly nicer than anywhere else I had ever seen at an airport, and perhaps in my life.

There were leather armchairs, fireplaces, marble bathrooms, a spa, a restaurant, a concierge, a manicurist and a hairdresser. One waiter toured the lounge with plates of complimentary caviar, foie gras and smoked salmon, while a second made circuits with éclairs and strawberry tartlets.

'For what purpose is all the toil and bustle of this world? What is the end of the pursuit of wealth, power and pre-eminence?' asked Adam Smith in *The Theory of Moral Sentiments* (1759), going on to answer, 'To be observed, to be attended to, to be taken notice of with sympathy, complacency, and approbation' – a set of ambitions to which the creators of the Concorde Room had responded with stirring precision.

As I took a seat in the restaurant, I felt certain that whatever it had taken for humanity to arrive at this point had ultimately been worth it. The development of the combustion engine, the invention of the telephone, the Second World War, the introduction of real-time financial information on Reuters screens, the Bay of Pigs, the extinction of the slender-billed curlew – all of these things had, each in its own fashion, helped to pave the way for a disparate group of uniformly attractive individuals to silently mingle in a splendid room with a view of a runway in a cloud-bedecked corner of the Western world.

'There is no document of culture that is not at the same time a document of barbarism,' the literary critic Walter Benjamin had once famously written, but that sentiment no longer seemed to matter very much.

Still, I recognised the fragility of the achievement behind the lounge. I sensed how relatively few such halcyon days there might be left before members of the small fraternity ensconced in its armchairs came to grief and its gilded ceilings cracked into ruin. Perhaps it had felt a bit like this on the terraces of Hadrian's villa outside Rome on autumn Sunday evenings in the second century AD, as a blood-red sun set over the marble colonnades. One might have had a similar presentiment of catastrophe, looming in the form of the restless Germanic tribes lying in wait deep in the sombre pine forests of the Rhine Valley.

I started to feel sad about the fact that I might not be returning to the Concorde Room anytime soon. I realised, however, that the best way to attenuate my grief would be to nurture a thoroughgoing hatred of all those more regularly admitted into the premises. Over a plate of porcini mushrooms on a brioche base, I therefore tried out the idea that the lounge was really a hideout for a network of oligarchs who had won undeserved access through varieties of nepotism and skulduggery.

Regrettably, on closer examination, I was forced to concede that the evidence conflicted unhelpfully with this otherwise consoling thesis, for my fellow guests fitted none of the stereotypes of the rich. Indeed, they stood out chiefly on the basis of how *ordinary* they looked. These were not the chinless heirs to hectares of countryside but rather normal people who had figured out how to make the microchip and spreadsheet work on their behalf. Casually dressed, reading books by Malcolm Gladwell, they were an elite who had come into their wealth by dint of intelligence and stamina. They worked at Accenture fixing irregularities in supply chains or built income-ratio models at MIT; they had started telecommunications companies or did astrophysical research at the Salk Institute. Our society is affluent in large part because its wealthiest citizens do not behave the way rich people are popularly supposed to. Simple plunder could never have built up this sort of lounge (globalised, diverse, rigorous, technologically-minded), but at best a few gilded pleasure palaces standing out in an otherwise feudal and backward landscape.

In the rarefied air that was pumped into the Concorde Room, there nonetheless hovered a hint of something troubling: the implicit suggestion that the three traditional airline classes represented nothing less than a tripartite division of society according to people's genuine talents and virtues. Having abolished the caste systems of old and fought to ensure universal access to education and opportunity, it seemed that we might have built up a meritocracy that had introduced an element of true justice into the distribution of wealth as well as of poverty. In the modern era, destitution could therefore be regarded as not merely pitiable but *deserved*. The question of why, if one was in any way talented or adept, one was still unable to earn admittance to an elegant lounge was a conundrum for all economy airline passengers to ponder in the privacy of their own minds as they perched on hard plastic chairs in the overcrowded and chaotic public waiting areas of the world's airports.

The West once had a powerful and forgiving explanation for exclusion from any sort of lounge: for two thousand years Christianity rejected the notion, inherent in the modern meritocratic system, that virtue must inevitably usher in material success. Jesus was the highest man, the most blessed, and yet throughout his earthly life he was poor, thus by his very example ruling out any direct equation between righteousness and wealth. The Christian story emphasised that, however apparently equitable our educational and commercial infrastructures might seem, random factors and accidents would always conspire to wreck any neat alignment between the hierarchies of wealth on the one hand and virtue on the other. According to St Augustine, only God himself knew what each individual was worth, and He would not reveal that assessment before the time of the Last Judgement, to the sound of thunder and the trumpets of angels – a phantasmagorical scenario for non-believers, but helpful nevertheless in reminding us to refrain from judging others on the basis of a casual look at their tax returns.

The Christian story has neither died out nor been forgotten. That it continues even now to scratch away at meritocratic explanations of privilege was made clear to me when, after a copious lunch rounded off by a piece of chocolate cake with passionfruit sorbet, an employee called Reggie described for me the complicated set of circumstances that had brought her to the brutally decorated staff area of the Concorde Room from a shantytown outside Puerto Princesa in the Philippines. Our preference for the meritocratic versus the Christian belief system will in the end determine how we decide to interpret the relative standing of a tracksuited twenty-seven-year-old entrepreneur reading the *Wall Street Journal* by a stone-effect fireplace while waiting to board his flight to Seattle, against that of a Filipina cleaner whose job it is to tour the bathrooms of an airline's first-class lounge, swabbing the shower cubicles of their diverse and ever-changing colonies of international bacteria.

6

Although the majority of its users regarded it as little more than a place where they had to spend a few hours on their way to somewhere else, for many others the terminal served as a permanent office, one that accommodated a thousand-strong bureaucracy across a series of floors off limits to the general public. The work done here was not well suited to those keen on seeing their own identities swiftly or flatteringly reflected back at them through their labour. The terminal had taken some twenty years and half a million people to build, and now that it was finally in operation, its business continued to proceed ponderously and only by committee. Layer upon layer of job titles (Operational Resource Planning Manager, Security Training and Standards Adviser, Senior HR Business Partner) gave an indication of the scale of the hierarchies that had to be consulted before a new computer screen could be acquired or a bench repositioned.

A few of the more obscure offices nonetheless managed to convey an impressive sense of the scope of the manpower and intelligence involved in getting planes around the world. The area housing British Airways' Customer Experience Division was filled with prototypes of cabin seats, life jackets, vomit bags, mints and towelettes. An archivist oversaw a room filled with rejected samples, most of which had ended up there on the grounds of cost, not so much because of the airline's miserliness as because of its sheer size, an overspend on a single chair having dramatic consequences when typing out a purchase order for thirty thousand of them. A tour of the premises, with a close look at the early designs for plane interiors, offered a pleasure similar to that of looking through the first draft of a manuscript and seeing that prose that would eventually be polished and sure had started out hesitant and confused – a lesson with consoling applications for a universal range of maiden efforts.

I came away from the back rooms of the airport regretting what seemed a wrong-headed imbalance between the lavish attention paid to distracting and entertaining travellers and the scant time spent educating them about the labour involved in their journeys.

It is a good deal more interesting to find out how an airline meal is made than how it tastes – and a good deal more troubling. A mile from the terminal, in a windowless refrigerated factory owned by the Swiss company Gate Gourmet, eighty thousand breakfasts, lunches and dinners, all intended for ingestion within the following fifteen hours somewhere in the troposphere, were being made up by a group of women from Bangladesh and the Baltic. Korean Airlines would be serving beef broth, JAL salmon teriyaki and Air France a chicken escalope on a bed of puréed carrots. Foods that would later be segregated according to airline and destination now mingled freely together, like passengers in the terminal, so that a tray containing a thousand plates of Dubai-bound Emirates hummus might be lined up in the freezer room next to four trolleys full of SAS gravadlax, set to fly part-way to Stockholm.

Aeroplane food stands at a point of maximum tension between the man-made and the natural, the technological and the organic. Even the most anaemic tomato (and the ones at Gate Gourmet were mesmerising in their fibrous pallor) remains a work of nature. How strange and terrifying, then, that we should take our fruit and vegetables up into the sky with us, when we used to sit more humbly at nature's feet, hosting harvest festivals to honour the year's wheat crop and sacrificing animals to ensure the continued fecundity of the earth.

There is no need for such prostration now. A batch of twenty thousand cutlets, which had once, if only briefly, been attached to lambs born and nursed on Welsh hillsides, was driven into the depot. Within hours, with the addition of a breadcrumb

topping, a portion of these would metamorphose into meals that would be eaten over Nigeria – with no thought or thanks given to their author, twenty-six-year-old Ruta from Lithuania.

7 The British Airways flight crews also maintained offices at the airport. In an operations room in Terminal 5, pilots stopped by throughout the day and into the evening to consult with their managers about what the weather was like over Mongolia, or how much fuel they ought to purchase in Rio. When I saw an opening, I introduced myself to Senior First Officer Mike Norcock, who had been flying for fifteen years and who greeted me with one of those wry, indulgent smiles often bestowed by professionals upon people with a more artistic calling. In his presence, I felt like a child unsure of his father's affections. I realised that meeting pilots was doomed to escalate into an ever more humiliating experience for me, as the older I got, the more obvious it became that I would never be able to acquire the virtues that I so admired in them – their steadfastness, courage, decisiveness, logic and relevance – and must instead forever remain a hesitant and inadequate creature who would almost certainly start weeping if asked to land a 777 amid foggy ground conditions in Newfoundland.

Norcock had come to the operations room to pick up some route maps. He was off to India in his jumbo but first wanted to double-check the weather over Iran's northern border. He knew so much that his passengers did not. He understood, for example, that the sky, which we laypeople so casually and naïvely tend to appraise in terms of its colour and cloud formations, was in fact criss-crossed by coded flight lanes, intersections, junctions and beacon signals. On this day, he was especially concerned with VAN115.2, both a small orange dot on the flight charts and a wooden shed two metres high and five across, situated on the edge of a farmer's field at the top of a gorge in a thinly inhabited part of eastern Turkey – a location where Norcock would in a few hours be taking a left fork on to airway R659, as his passengers anxiously anticipated their lunch, a lasagne being prepared even now in Gate Gourmet's factory. I looked at his steady, well-sculpted hands and thought of how far he had come since childhood.

I knew, at least in theory, that Norcock could not always, in every circumstance, be a model of authoritative and patriarchal behaviour. He, too, must be capable of petulance, of vanity, of acting foolishly, of making casually cruel remarks to his spouse or neglecting to understand his children. There are no directional charts for daily life. But at the same time, I was reluctant to either accept or exploit the implications of this knowledge. I wanted to believe in the capacity of certain professions to enable us to escape the ordinary run of our frailties and to accede, if only for a moment, to a more impressive sort of existence than most of us will ever know.

8

From the outset, my employer had suggested that I might wish to conduct a brief interview with one of the most powerful men in the terminal: the head of British Airways, Willie Walsh. It was a daunting prospect, as Walsh was having a busy time of it. His company was losing an average of £1.6 million a day, a total of £148 million over the previous three months. His pilots and cabin crew were planning strikes. Studies showed that his baggage handlers misappropriated more luggage than their counterparts at any other European airline. The government wanted to tax his fuel. Environmental activists had been chaining themselves to his fences. He had infuriated those in the upper echelons at Boeing by telling them that he would not be able to keep up with the prepayment schedule he had committed to for the new 787 aircraft he had ordered. His efforts to merge his airline with Qantas and Iberia had stalled. He had done away with the free chocolates handed round after every meal in business class, and in the process provoked a three-day furore in the British press.

Journalism has long been enamoured of the idea of the interview, beneath which lies a fantasy about access: a remote figure, beyond the reach of the ordinary public and otherwise occupied with running the world, opens up and reveals his innermost self to a correspondent. With admission set at the price of a newspaper, the audience is invited to forget their station in life and accompany the interviewer into the palace or the executive suite. The guards lay down their weapons, the secretaries wave the visitors through. Now we are in the inner sanctum. While waiting, we have a look around. We learn that the president likes to keep a bowl of peppermints on his desk, or that the leading actress has been reading Dickens.

But the tantalising promise of shared secrets is rarely fulfilled as we might wish, for it is almost never in the interests of a prominent figure to become intimate with a member of the press. He has better people on to whom to unburden himself. He does not need a new friend. He is not going to disclose his plots for vengeance or his fears about his professional future. For the

celebrity, the interview is thus generally reduced to an exercise in saying as little as possible without confounding the self-love of the journalist on the sofa, who might become dangerous if rendered too starkly aware of the futility of his mission. In a bid to appease the underlying demand for closeness, the subject may let it drop that he is about to go on holiday to Florida, or that his daughter is learning how to play tennis.

There was evidently nothing of standard consequence that I could ask Mr Walsh. There was no point in my bringing up pensions, carbon emissions, premium yields or even the much-missed chocolates – no point, really, in our meeting at all, had not events reached the stage where articulating this insight would have seemed rude.

So we got together for forty minutes in a conference room, between Mr Walsh's meeting with a trade-union representative and a delegation from Airbus. I felt as if I were interrupting a discussion of beachheads between Roosevelt and Churchill in May 1943.

Fortunately, I had come to the conclusion that though Mr Walsh was the CEO of one of the world's largest airlines, it would be wholly unfair of me to treat him like a businessman.

The fiscal state of his company was simply too precarious, and too woefully inaccurate a reflection of his talents and interests, to permit me to confuse Mr Walsh with his balance sheet.

Considered collectively, as a cohesive industry, civil aviation had never in its history shown a profit. Just as significantly, neither had book publishing. In this sense, then, the CEO and I, despite our apparent differences, were in much the same sort of business, each one needing to justify itself in the eyes of humanity not so much by its bottom line as by its ability to stir the soul. It seemed as unfair to evaluate an airline according to its profit-and-loss statement as to judge a poet by her royalty statements. The stock market could never put an accurate price on the thousands of moments of beauty and interest that occurred around the world every day under an airline's banner: it could not describe the sight of Nova Scotia from the air, it had no room in its optics for the camaraderie enjoyed by employees in the Hong Kong ticket office, it had no means of quantifying the adrenalin-thrill of take-off.

The logic of my argument was not lost on Mr Walsh, who had himself once been a pilot. As we talked, he expressed his admiration for the way planes, vast and complicated machines, could defy their size and the challenges of the atmosphere to soar into the sky. We remarked on the surprise we both felt on seeing a 747 at a gate, dwarfing luggage carts and mechanics, at the idea that such a leviathan could move – a few metres' distance, let alone across the Himalayas. We reflected on the pleasure of seeing a 777 take off for New York and, over the Staines reservoir, retract its flaps and wheels, which it would not require again until its descent over the white clapboard houses of Long Beach, some 5,000 kilometres and six hours of sea and cloud away. We exclaimed over the beauty of a crowded airfield, where, through the heat haze of turbofans, the interested observer can make out sequences of planes waiting to begin their journeys, their fins a confusion of colours against the grey horizon, like sails at a regatta. In another life, I decided, the chief executive and I might have become good friends.

We were getting on so well that Mr Walsh – or Willie, as he now urged me to call him – suggested we repair to the lobby downstairs, where we could have a look at a model of the new A380, twelve of which he had ordered from Airbus and which would be joining the British Airways fleet in 2012. Once we were standing before it, Willie, with what seemed a child's sense of delight, invited me to join him in climbing up on to a bench to appreciate the sheer scale of the jet's ailerons and the breadth of its fuselage.

So much warmth did I feel for him as we stood shoulder to shoulder, admiring his model plane, that I was emboldened to mention a fantasy I had harboured since I first received authorisation to write a book about Heathrow. I asked Willie whether, if he had any money left, he might one day consider appointing me his writer-in-flight, in order that I might constantly circumnavigate the earth composing, among other things, sincere dedications to my patron, impressionistic essays describing the ochre colours of the western Australian desert as seen from the flight deck, and vignettes recounting the balletic routines of the stewards in the galley.

There was a pause, and for a moment the bonhomie disappeared from the chief executive's handsome grey-green eyes. But soon enough it returned. 'Of course,' he said, beaming. 'Once at Aer Lingus, the video system broke down, and we invited a couple of Irish minstrels to sing songs on a flight to New York. Alan, I could see you at the front of the cabin doing a ditty or two for our passengers.' And following that prognostication, he apologised for taking up so much of my precious time and called for a security officer to escort me to the door of his corporate headquarters.

9

Not long into my stay, evening became my favourite time at the airport. By eight, most of the choppy short-haul European traffic had come and gone. The terminal was emptying out, Caviar House was selling the last of its sturgeon eggs and the cleaning teams were embarking on the day's most systematic mopping of the floors. Because it was summer, the sun would not set for another forty minutes, and in the interim a gentle, nostalgic light would flood across the seating areas.

The majority of the passengers left in the terminal at this hour were booked on one or another of the flights that departed every evening for the East, unbeknownst to most of the households of north-west London which they crossed en route for Singapore, Seoul, Hong Kong, Shanghai, Tokyo and Bangkok.

The atmosphere in the waiting areas was lonely, but curiously, the feeling was benign for being so general, eliminating the unease that any one individual might otherwise experience at being the only one to be alone, and thus paradoxically making new connections seem possible in a way they might not have done in the more obviously convivial surroundings of a crowded city bar. At right, the airport emerged as a home of nomadic spirits, types who could not commit to any one country, who shied from tradition and were suspicious of settled community, and who were therefore nowhere more comfortable than in the intermediate zones of the modern world, landscapes gashed by kerosene storage tanks, business parks and airport hotels.

Because the arrival of night typically pulls us back towards the hearth, there seemed something especially brave about travellers who were preparing to entrust themselves to the darkness, to be carried in a craft navigated by instruments alone and to surrender to sleep, finally, only over Azerbaijan or the Kalahari Desert.

In a control room beside the terminal, a giant map of the world showed the real-time position of every plane in the British Airways fleet, as tracked by a string of satellites. Across the globe, 180 aircraft were on their way, together holding some one hundred thousand passengers. A dozen planes were crossing the North Atlantic, five were routing around a hurricane to the west of Bermuda, and one could be seen plotting a course over Papua New Guinea. The map was emblematic of a touching vigilance, for however far removed each craft was from its home

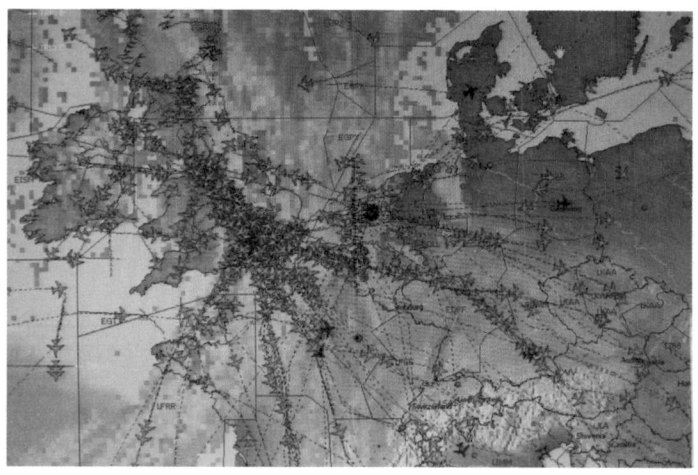

airfield, however untethered and able it looked, it was never far from the minds of those in the control room in London, who, like parents worrying about their children, would not feel at ease until each of their charges had safely touched down.

Every night a few planes would be towed away from their gates to a set of giant hangars, where a phalanx of gangways and cranes would lock themselves around their organically shaped bodies like a series of handcuffs. While aircraft tended to be coy about their need to pay such visits – hardly letting on, at the close of a trip from Los Angeles or Hong Kong, that they had reached the very end of their permitted quota of nine hundred flying hours – the checks provided an opportunity for them to reveal their individuality. What to passengers might have looked like yet another indistinguishable 747 would emerge, during this process, as a machine with a distinct name and medical history: G-BNLH, for example, had come into service in 1990 and in the intervening years had had three hydraulic leaks over the Atlantic, once blown a tyre in San Francisco and, only the previous week, dropped an apparently unimportant part of its wing in Cape Town. Now it was coming into the hangar with, among other ailments, twelve malfunctioning seats, a large smear of purple nail polish on a wall panel and an opinionated

microwave oven in a rear galley that ignited itself whenever an adjacent basin was used.

Thirty men would work on the plane through the night, the whole operation guided by an awareness that, while the craft could under most circumstances be extraordinarily forgiving, a chain of events originating in the failure of something as small as a single valve could nevertheless bring it down, just as a career might be ruined by one incautious remark, or a person die because of a clot less than a millimetre across.

I toured the exterior of the aircraft on a gangway that ran around its midriff and let my hands linger on its nose cone, which had only a few hours earlier carved a path through dense layers of static cumulus clouds.

Studying the plane's tapered tail, and the marks left across the back of its fuselage by the enraged thrust of its four RB211 engines, I wondered if scientists and engineers might have designed planes and their means of take-off differently had our species been graced with some subtler, less thunderous mode of conception, perhaps one managed frictionlessly and quietly by the male's sitting for a few hours on an egg left behind in a leafy recess by the female.

10 At around eleven-fifteen each night, by government decree, the airport was closed to both incoming and outgoing traffic. Across the aprons, all was suddenly as quiet as it must have been a hundred years ago, when there was nothing here but sheep meadows and apple farms. I met up with a man called Terry, whose job it was to tour the runways in the early hours looking out for stray bits of metal. We drove out to a spot at the end of the southern runway, 27L to pilots, which Terry termed the most expensive piece of real estate in Europe. It was here, at forty-second intervals throughout the day, on a patch of tarmac only a few metres square and black with rubber left by tyres, that the aircraft of the world made their first contact with the British Isles. This was the exact set of coordinates that planes anticipated from across southern England: even in the thickest fog, their automatic landing systems could pick up the glide-path beam that was projected up into the sky from this point, the radio wave calling them to place their wheels squarely in the centre of a zone highlighted by a double line of parallel white lights.

But just now, the patch of runway that was almost solely responsible for destroying the peace and quiet of some ten million people was becalmed. One could walk unhurriedly across it and even give in to the temptation to sit cross-legged on its centreline, a gesture that partook of some of the sublime thrill of touching a disconnected high-voltage electricity cable, running one's fingers along the teeth of an anaesthetised shark or having a wash in a fallen dictator's marble bathroom.

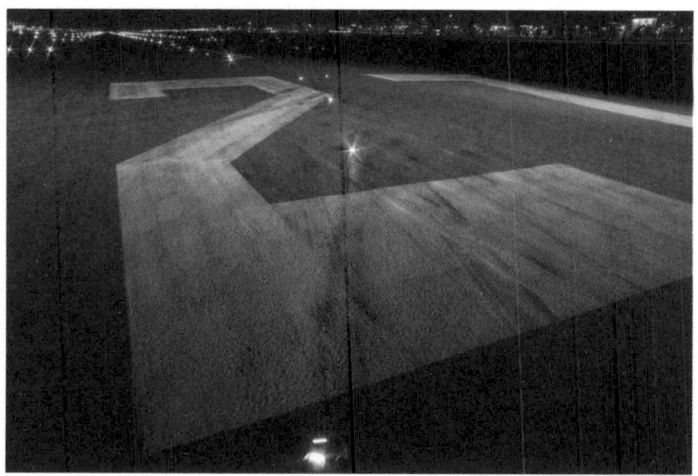

A field mouse scurried out of the grass and on to the runway, where for a moment it stood still, transfixed by the jeep's headlamps. It was of a kind which regularly populates children's books, where mice are always clever and good-natured creatures who live in small houses with red-and-white-checked curtains, in sharp contrast to the boorish humans, who are clumsily oversized and unaware of their own limits. Its presence this night on the moonlit tarmac served optimistically to suggest that when mankind is finished with flying – or more generally, with *being* – the earth will retain a capacity to absorb our follies and make way for more modest forms of life.

11 Terry dropped me off at my hotel. I felt too stimulated to sleep – and so went for a drink at an all-night bar frequented by delayed flight crews and passengers.

Over a dramatically sized tequila-based cocktail named an After Burner, I befriended a young woman who told me that she was writing a doctoral dissertation at the University of Warsaw: her subject was the Polish poet and novelist Zygmunt Krasiński, with a particular focus on his famous work *Agaj-Han* (1834) and the tragic themes explored therein. She argued that Krasiński's reputation had been unfairly eclipsed during the twentieth century by that of his fellow Romantic writer Adam Mickiewicz, and explained that she had been motivated in her research by a desire to reacquaint her compatriots with an aspect of their heritage that had been deliberately denied them during the Communist era. When I asked why she was at the airport, she replied that she had come to meet a friend from Dubai, whose plane had been delayed and was now unlikely to land at Heathrow before mid-morning. An engineer of Lebanese origin, he had been coming to London once a month for the past year and a half in order to receive treatment for throat cancer at a private hospital in Marylebone, and during every visit, he invited her to spend the night with him in a Prestige

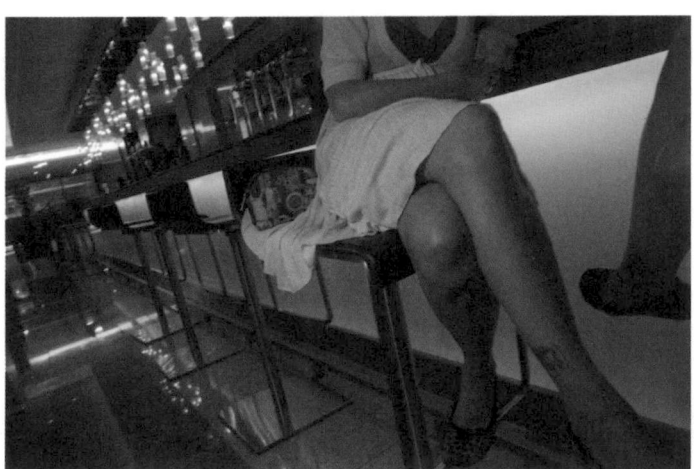

Suite on the top floor of the Sofitel. She confided that she was registered with an agency which had a head office in Hayes and added, in a not unrelated aside, that Zygmunt Krasiński had conducted a three-year-long affair with the Countess Delfina Potocka, with whom Chopin had also been in love.

I returned to my room at three in the morning, struck by a sense of our race as a peculiar, combustible mixture of the beast and the angel. The first plane, due at Heathrow at dawn, was now somewhere over western Russia.

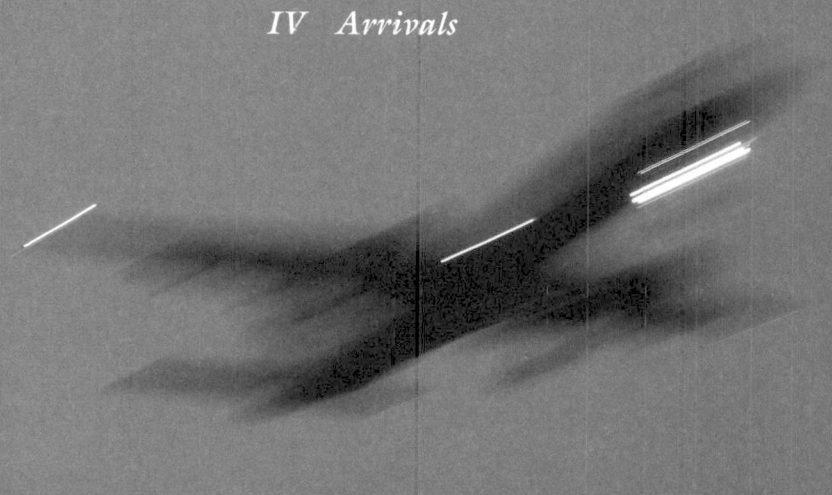

IV Arrivals

1 There used to be time to arrive. Incremental geographical changes would ease the inner transitions: desert would gradually give way to shrub, savannah to grassland. At the harbour, the camels would be unloaded, a room would be found overlooking the customs house, passage would be negotiated on a steamer. Flying fish would skim past the ship's hull. The crew would play cards. The air would cool.

Now a traveller may be in Abuja on Tuesday and at the end of a satellite in the new terminal at Heathrow on Wednesday. Yesterday lunchtime, one had fried plantain in the Wuse District to the sound of an African cuckoo, whereas at eight this morning the captain is closing down the 777's twin engines at a gate next to a branch of Costa Coffee.

Despite one's exhaustion, one's senses are fully awake, registering everything – the light, the signage, the floor polish, the skin tones, the metallic sounds, the advertisements – as sharply as if one were on drugs, or a newborn baby, or Tolstoy. Home all at once seems the strangest of destinations, its every detail relativised by the other lands one has visited. How peculiar this morning light looks against the memory of dawn in the Obudu hills, how unusual the recorded announcements sound after the wind in the High Atlas and how inexplicably English

(in a way they will never know) the chat of the two female ground staff seems when one has the din of a street market in Lusaka still in one's ears.

One wants never to give up this crystalline perspective. One wants to keep counterpoising home with what one knows of alternative realities, as they exist in Tunis or Hyderabad. One wants never to forget that nothing here is normal, that the streets are different in Wiesbaden and Luoyang, that this is just one of many possible worlds.

2 In the brief history of aviation, not many airports have managed to fulfil their visitors' hopes for an architecture that might properly honour the act of arrival. Too few have followed the example set by Jerusalem's elaborate Jaffa Gate, which once welcomed travellers who had completed the journey to the Holy City across the baked Shephelah plains and through the thief-infested Judean hills. But Terminal 5 wanted to have a go.

In the older terminals at Heathrow, it was a certain sort of carpet that one tended to notice first, swirling green, yellow, brown and orange, around which there hovered associations of vomit, pubs and hospitals. Here, by way of contrast, there were handsome grey composite tiles, bright corridors lined with glass panels in a calming celadon shade and bathrooms fitted out with gracious sanitary ware and full-length cubicle doors made of heavy timber.

The structure was proposing a new idea of Britain, a country that would be reconciled to technology, that would no longer be in thrall to its past, that would be democratic, tolerant, intelligent, playful and lacking in spite or irony. All this was a simplification, of course: twenty kilometres to the west and north were tidy hamlets and run-down estates that would at once have contravened any of the suggestions encoded in the terminal's walls and ceilings.

Nevertheless, like Geoffrey Bawa's Parliament in Colombo or Jørn Utzon's Opera House in Sydney, Richard Rogers's Terminal 5 was applying the prerogative of all ambitious architecture to create rather than merely reflect an identity. It hoped to use the hour or so when passengers were within its space – objectively, to have their passports stamped and to recover their luggage – to define what the United Kingdom might one day become, rather than what it too often is.

3   Upon disembarking, after a short walk, arriving passengers entered a hall that tried hard to downplay the full weight of its judicial role. There were no barriers, guns or reinforced booths, merely an illuminated sign overhead and a thin line of granite running across the floor. Power was sure of itself here, confident enough to be restrained and invisible to those privileged, by an accident of birth, to skirt it. Three times a day, a cleaning team came and swept their brooms across the line that marked the divide between the no man's land of the aircraft on the one side and, on the other, the well-stocked pharmacies, benign mosquitoes, generous library lending policies, sewage plants and pelican crossings available to visitors and residents of Great Britain alike.

With just a single unhappy swipe of the computer, however, all such implicit promises might be prematurely broken. A guard would be called and would lead the unfortunate traveller from the immigration hall to a suite of rooms two storeys below. The children's playroom seemed especially poignant in its fittings: there was a Brio train, most of the Lego City range, a box of Caran d'Ache pens and, for each new child sequestered there, a box of snacks and plastic animals, his or hers to keep.

In the imaginations of certain children in Eritrea or Somalia, England would hence always remain a briefly glimpsed country of Quavers, Jelly Tots and squared cartons of orange juice – a country so rich it could afford to give away small digital alarm clocks, and one whose guards knew how to put wooden train tracks together. Next door, in a barer room in which every word was being captured by a police tape recorder, their parents would experience another side of the nation, as they delineated their unsuccessful applications to an impassive member of the immigration service.

4 Over the course of history, few joyful moments can have unfolded in a baggage-reclaim area, though the one in the terminal was certainly doing its best to keep its users optimistic.

It had high ceilings, flawlessly poured concrete walls and trolleys in abundance. Furthermore, the bags came quite quickly. The company responsible for the conveyor belts, Vanderlande Industries from the Netherlands, had made its reputation in the mail-order and parcel-distribution sectors and was now the world leader in suitcase logistics. Seventeen kilometres' worth of conveyor belts ran under the terminal, where they were capable of processing some twelve thousand pieces of luggage an hour. One hundred and forty computers scanned tags, determined where individual bags were going and checked them for explosives along the way. The machines treated the suitcases with a level of care that few humans would have shown them: when the bags had to wait in transit, robots would carry them gently over to a dormitory and lay them down on yellow mattresses, where – like their owners in the lounges above – they would loll until their flights were ready to receive them. By the time they were lifted off the belt, many suitcases were likely to have had more interesting travels than their owners.

Nevertheless, in the end, there was something irremediably melancholic about the business of being reunited with one's luggage. After hours in the air free of encumbrance, spurred on to formulate hopeful plans for the future by the views of coasts and forests below, passengers were reminded, on standing at the carousel, of all that was material and burdensome in existence. There were some elemental dualities at work in the contrasting realms of the baggage-reclaim hall and the aeroplane – dichotomies of matter and spirit, heaviness and lightness, body and soul – with the negative halves of the equations all linked to the stream of almost identical black Samsonite cases that rolled ceaselessly along the tunnels and belts of Vanderlande's exquisite conveyor apparatus.

Around the carousel, as in a Roman traffic jam, trolleys grimly refused to cede so much as a centimetre to one another. Although each suitcase was a repository of dense and likely fascinating individuality – this one perhaps containing a lime-coloured bikini and an unread copy of *Civilization and Its Discontents,* that one a dressing gown stolen from a Chicago hotel and a packet of Roche antidepressants – this was not the place to start thinking about anyone else.

5 Yet the baggage area was only a prelude to the airport's emotional climax. There is no one, however lonely or isolated, however pessimistic about the human race, however preoccupied with the payroll, who does not in the end expect that someone significant will come to say hello at arrivals.

Even if our loved ones have assured us that they will be busy at work, even if they told us they hated us for going travelling in the first place, even if they left us last June or died twelve and a half years ago, it is impossible not to experience a shiver of a sense that they may have come along anyway, just to surprise us and make us feel special (as someone must have done for us when we were small, if only occasionally, or we would never have had the strength to make it this far).

It is therefore hard to know just what expression we should mould our faces into as we advance towards the reception zone. It might be foolhardy to relinquish the solemn and guarded demeneaour we usually adopt while wandering through the anonymous spaces of the world, but at the same time, it seems only right that we should leave open at least the suggestion of a smile. We may settle on the sort of cheerful but equivocal look commonly worn by people listening out for punchlines to jokes narrated by their bosses.

So what dignity must we possess not to show any hesitation when it becomes clear, in the course of a twelve-second scan of the line, that we are indeed alone on the planet, with nowhere to head to other than a long queue at the ticket machine for the Heathrow Express. What maturity not to mind that only two metres from us, a casually dressed young man perhaps employed in the lifeguard industry has been met with a paroxysm of joy by a sincere and thoughtful-looking young woman with whose mouth he is now involved. And what a commitment to reality it will take for us not to wish that we might, just for a time, be not our own tiresome selves but rather Gavin, flying in from

Los Angeles after a gap year in Fiji and Australia, with whose devoted parents, exhilarated aunt, delighted sister, two girl friends and a helium balloon we might therefore repair to a house on the southern outskirts of Birmingham.

At arrivals, there were forms of welcome of which princes would have been jealous, and which would have rendered inadequate the celebrations laid on at Venice's quaysides for the explorers of the Eastern silk routes. Individuals without official status or distinguishing traits, passengers who had sat unobtrusively for twenty-two hours near the emergency exits, now set aside their bashfulness and revealed themselves as the intended targets of flags, banners, streamers and irregularly formed home-baked chocolate biscuits – while, behind them, the chiefs of large corporations prepared for glacial limousine rides to the marble-and-orchid-bedecked lobbies of their luxury hotels.

The prevalence of divorce in modern society guaranteed an unceasing supply of airport reunions between parents and children. In this context, there was no longer any point in pretending to be sober or stoic: it was time to squeeze a pair of frail and yet plump shoulders very tightly and founder into tears. We may spend the better part of our professional lives projecting strength and toughness, but we are all in the end creatures of appalling fragility and vulnerability. Out of the millions of people we live among, most of whom we habitually ignore and are ignored by in turn, there are always a few who hold hostage our capacity for happiness, whom we could recognise by their smell alone and whom we would rather die than be without. There were men pacing impatiently and blankly who had looked forward to this moment for half a year and could not restrain themselves any further at the sight of a small boy endowed with their own grey-green eyes and their mother's cheeks, emerging from behind the stainless-steel gate, holding the hand of an airport operative.

At such moments, it felt almost as if death itself had been averted – and yet there was also a sense, lending the occasion more poignancy still, that it could not go on being cheated for ever. Perhaps this was a way of practising for mortality. Some day, many years from now, the adult child would say goodbye to his father before going on a routine business trip, and the reprieve would abruptly run out. There would be a telephone call in the middle of the night to a room on the twentieth floor of a Melbourne hotel, bringing the news that the parent had suffered a catastrophic seizure on the other side of the world and that there was nothing more the doctors could do for him – and from that day forward, for the now-grown-up boy, the line in arrivals would always be missing one face in particular.

6 Not all meetings were so emotional. One might have come from Shanghai to join Malcolm and Mike for a drive down to Bournemouth to learn English for the summer: a two-month sojourn in a bed and breakfast near the pier, with regular lessons from a tutor who would teach her class how to say 'ought' and help them master business English, a subcategory of the language that would vouchsafe future careers in the semi-conductor and textile industries of the Pearl River delta.

For his part, Mohammed was waiting for Chris's flight from San Francisco. The former, originally from Lahore, was at present based in Southall, while the latter, from Portland, Oregon, now lived in Silicon Valley – not that either man would attempt to discover these details about the other. In an otherwise uninhabited universe, how strange that one should so easily be able to sit in silence with another human being in a black Mercedes S-Class sedan. For both driver and passenger, the trip would be counted a success if the other party proved not to be a murderer or a thief. The hour and a half of stillness would be punctuated only by the occasional electronic command to turn left or right at the next junction, until the Mercedes reached a glass-fronted office building in Canary Wharf, where Chris was due to attend a meeting on the storage of financial data and Mohammed returned to the terminal to begin another journey, this time to Kent, with the no less mysterious or more talkative Mr K from Narita.

7 Many of the more conventional reunions seemed to beg the question of how their levels of excitement could be kept up. Maya had been waiting for this moment for the previous twelve hours. She had had butterflies since her plane crossed the coast of Ireland. At 9,000 metres up, she had anticipated Gianfranco's touch. But finally, after eight minutes of a sustained embrace, the couple had no alternative: it was time for them to go and find his car.

It seems curious but in the end appropriate that life should often put in our way, so near to the site of some of our most intense and heartfelt encounters, one of the greatest obstacles known to relationships: the requirement to pay for and then negotiate a way out of a multi-storey car park.

Then again, as we strain to remain civil under the unforgiving fluorescent lights, we may be reminded of one of the reasons we went travelling in the first place: to make sure that we would be better able to resist the mundane and angry moods in which daily life is so ready to embroil us.

The very brutality of the setting – the concrete floor marred with tyre marks and oil stains, the bays littered with abandoned trolleys and the ceilings echoing to the argumentative sounds of slamming doors and accelerating vehicles – encourages us to steel ourselves against a slide back into our worst possibilities. We may ask of our destinations, 'Help me to feel more generous, less afraid, always curious. Put a gap between me and my confusion; the whole of the Atlantic between me and my shame.' Travel agents would be wiser to ask us what we hope to change about our lives rather than simply where we wish to go.

The notion of the journey as a harbinger of resolution was once an essential element of the religious pilgrimage, defined as an excursion through the outer world undertaken in an effort to promote and reinforce an inner evolution. Christian theorists were not in the least troubled by the dangers, discomforts or expense posed by pilgrimages, for they regarded these and other apparent disadvantages as mechanisms whereby the

underlying spiritual intent of the trip could be rendered more vivid. Snowbound passes in the Alps, storms off the coast of Italy, brigands in Malta, corrupt Ottoman guards – all such trials merely helped to ensure that a trip would not be easily forgotten.

Whatever the benefits of prolific and convenient air travel, we may curse it for its smooth subversion of our attempts to use journeys to make lasting changes in our lives.

8   It was time to start packing up. In the connecting corridor leading to the Sofitel, I was intercepted by a fellow employee of the airport who was conducting a survey of newly arrived passengers, gathering their impressions of the terminal, from the signage to the lighting, the eating to the passport stamping. The responses were calibrated on a scale of 0 to 5, and the results would be tabulated as part of an internal review commissioned by the chief executive of Heathrow. I questioned the unusually protracted nature of this interview only in so far as it made me think of how seldom market researchers, with access to influential authorities, ask us to reflect on any of the more troubling issues we face in life more generally. On a scale of 0 to 5, how are we enjoying our marriages? Feeling about our careers? Dealing with the idea of one day dying?

I ordered my last commercially sponsored club sandwich in the hotel lounge. The planes were particularly loud as they passed overhead – so loud that as one MEA Airbus took off for Beirut, the waiter shouted, 'God help us!' in a way that startled me and my sole fellow diner, a businessman from Bangladesh en route to Canada.

I worried that I might never have another reason to leave the house. I felt how hard it is for writers to look beyond domestic experience. I dreamt of other possible residencies in institutions central to modern life – banks, nuclear power stations, governments, old people's homes – and of a kind of writing that could report on the world while still remaining irresponsible, subjective and a bit peculiar.

9   Just as passengers were concluding their journeys in the arrivals hall, above them, in departures, others were preparing to set off anew. BA138 from Mumbai was turning into BA295 to Chicago. Members of the crew were dispersing: the captain was driving to Hampshire, the chief purser was on a train to Bristol and the steward who had looked after the upper deck was already out of uniform (and humbled thereby, like a soldier without his regimental kit) and headed for a flat in Reading.

Travellers would soon start to forget their journeys. They would be back in the office, where they would have to compress a continent into a few sentences. They would have their first arguments with spouses and children. They would look at an English landscape and think nothing of it. They would forget the cicadas and the hopes they had conceived together on their last day in the Peloponnese.

But before long, they would start to grow curious once more about Dubrovnik and Prague, and regain their innocence with regard to the power of beaches and medieval streets. They would have fresh thoughts about renting a villa somewhere next year.

We forget everything: the books we read, the temples of Japan, the tombs of Luxor, the airline queues, our own foolishness. And so we gradually return to identifying happiness with elsewhere: twin rooms overlooking a harbour, a hilltop church boasting the remains of the Sicilian martyr St Agatha, a palm-fringed bungalow with complimentary evening buffet service. We recover an appetite for packing, hoping and screaming. We will need to go back and learn the important lessons of the airport all over again soon.

Acknowledgements

With many thanks to the following. At Mischief, Dan Glover (who had the idea), Charlotte Hutley and Seb Dilleyston. At BAA, Colin Matthews, Cat Jordan, Claire Lovelady and the Communications team, and Mike Brown and the Operations team. At Heathrow, Sofitel, British Airways, Gate Gourmet, the UK Border Authority and OCS. At Profile Books, Daniel Crewe, Ruth Killick and Paul Forty. Lesley Levene, Dorothy Straight and Fiona Screen for copy editing and proofreading. Richard Baker for the superlative images. Joana Niemeyer and David Pearson for the design. Caroline Dawnay and Nicole Aragi for the piloting. Charlotte, Samuel and Saul for another ruined August. In the text, some of the names have been changed to protect identities.